Renewable Energy Roadmap

A Step-by-Step Guide to Sustainable Power

Lane Becker

Renewable Energy Roadmap

*© Copyright 2024 by **Lane Becker***

All rights reserved

This document is geared towards providing exact and reliable information with regards to the topic and issue covered. The publication is sold with the idea that the publisher is not required to render accounting, officially permitted, or otherwise, qualified services. If advice is necessary, legal or professional, a practiced individual in the profession should be ordered.

From a Declaration of Principles which was accepted and approved equally by a Committee of the American Bar Association and a Committee of Publishers and Associations.

In no way is it legal to reproduce, duplicate, or transmit any part of this document in either electronic means or in printed format. Recording of this publication is strictly prohibited and any storage of this document is not allowed unless with written permission from the publisher. All rights reserved.

The information provided herein is stated to be truthful and consistent, in that any liability, in terms of inattention or otherwise, by any usage or abuse of any policies, processes, or directions contained within is the solitary and utter responsibility of the recipient reader. Under no circumstances will any legal responsibility or blame be held against the publisher for any reparation, damages, or monetary loss due to the information herein, either directly or indirectly.

Respective authors own all copyrights not held by the publisher.

The information herein is offered for informational purposes solely, and is universal as so. The presentation of the information is without contract or any type of guarantee assurance.

The trademarks that are used are without any consent, and the publication of the trademark is without permission or backing by the trademark owner. All trademarks and brands within this book are for clarifying purposes only and are owned by the owners themselves, not affiliated with this document.

TABLE OF CONTENTS

Types of Geothermal Systems 74

Innovations and Future Directions.. 86

Basics of Solar Power Technology

Solar power technology has emerged as a cornerstone of the renewable energy landscape, offering a sustainable solution to the world's growing energy demands. At its core, solar power harnesses the energy of the sun, converting sunlight into electricity through a process that is both efficient and environmentally friendly. The journey of solar power begins with the understanding of photovoltaic (PV) cells, the fundamental building blocks of solar panels. These cells are composed of semiconductor materials, typically silicon, which exhibit the photovoltaic effect—a phenomenon where light energy is converted into electrical energy.

The operation of a solar cell is rooted in the interaction between sunlight and the semiconductor material. When photons from sunlight strike the surface of a solar cell, they transfer their energy to electrons within the semiconductor. This energy boost allows the electrons to break free from their atomic bonds, creating electron-hole pairs. The movement of these free electrons generates an electric current, which can be harnessed for power. This process is facilitated by the presence of an electric field within the solar cell, which directs the flow of electrons, ensuring a consistent and usable current.

Solar panels, or modules, are composed of multiple solar cells connected in series and parallel configurations to achieve the desired voltage and current levels. The efficiency of a solar panel is determined by its ability to convert sunlight into electricity, with current commercial panels achieving efficiencies ranging from 15% to 22%. Advances in technology continue to push these boundaries, with research focused on developing new materials and cell designs that promise even higher efficiencies.

One of the key advantages of solar power technology is its versatility. Solar panels can be deployed in a variety of settings, from residential rooftops to large-scale solar farms. This adaptability allows for the integration of solar power into existing infrastructure, providing a decentralized energy solution that can reduce reliance on traditional power grids. Additionally, solar power systems can be paired with energy storage solutions, such as batteries, to provide a reliable and continuous power supply even when the sun is not shining.

The installation of solar power systems involves several critical steps, beginning with site assessment and system design. A thorough evaluation of the site is essential to determine the optimal placement and orientation of solar panels, maximizing exposure to sunlight and ensuring efficient energy production. Factors such as shading, roof angle, and local climate conditions must be considered to tailor the system to the specific needs of the location.

Once the design is finalized, the installation process involves mounting the solar panels, connecting the electrical components, and integrating the system with the existing electrical infrastructure. Proper installation is crucial to ensure the safety and performance of the solar power system. Regular maintenance, including cleaning and inspection of the panels and electrical components, is necessary to maintain optimal performance and extend the lifespan of the system.

The economic and environmental impacts of solar power technology are significant. On the economic front, the cost of solar panels has decreased dramatically over the past decade, making solar power an increasingly affordable option for both individuals and businesses. The initial investment in solar power systems can be offset by government incentives, tax credits, and the long-term savings on energy bills. Moreover, solar power systems can increase property values and provide a hedge against rising energy costs.

Environmentally, solar power offers a clean and renewable energy source that reduces greenhouse gas emissions and dependence on fossil fuels. By generating electricity without combustion, solar power minimizes air pollution and contributes to the mitigation of climate change. The lifecycle of solar panels, from production to disposal, is also being optimized to reduce environmental impact, with efforts focused on recycling and sustainable manufacturing practices.

The success of solar power technology is exemplified by numerous case studies around the world. In Germany, the Energiewende initiative has led to widespread adoption of solar power, contributing to the country's transition to a low-carbon energy system. In California, the Solar Star project, one of the largest solar farms in the world, generates enough electricity to power hundreds of thousands of homes, demonstrating the scalability of solar power solutions.

As solar power technology continues to evolve, innovations such as bifacial panels, which capture sunlight on both sides, and perovskite solar cells, which offer high efficiency at a lower cost, are poised to revolutionize the industry. These advancements promise to enhance the performance and accessibility of solar power, paving the way for a sustainable energy future.

In conclusion, the basics of solar power technology encompass a complex interplay of scientific principles, engineering practices, and environmental considerations. By harnessing the power of the sun, solar technology offers a viable path toward a cleaner and more sustainable energy landscape. As the world continues to embrace renewable energy solutions, solar power stands at the forefront, driving innovation and progress in the quest for a sustainable future.

Types of Solar Panels and Their Applications

Solar panels have become a ubiquitous symbol of renewable energy, capturing the sun's rays and converting them into electricity. However, not all solar panels are created equal. Different types of solar panels are designed to meet various needs and applications, each with its own set of advantages and limitations. Understanding these differences is crucial for selecting the right solar technology for specific projects, whether they are residential, commercial, or industrial.

Monocrystalline solar panels are among the most efficient and widely used types of solar panels. They are made from a single, continuous crystal structure, which gives them a uniform appearance and high efficiency rates, often exceeding 20%. The manufacturing process involves slicing thin wafers from a cylindrical silicon ingot, which is then treated to enhance its photovoltaic properties. Monocrystalline panels are known for their durability and long lifespan, making them an excellent choice for installations where space is limited, and maximum efficiency is desired. Their sleek, black appearance also makes them aesthetically pleasing, a factor that can be important for residential installations.

Polycrystalline solar panels, on the other hand, are made from multiple silicon crystals melted together. This process is less expensive than the one used for monocrystalline panels, resulting in a lower cost per watt. Polycrystalline panels typically have a blue hue and a slightly lower efficiency, usually ranging from 15% to 17%. Despite their lower efficiency, they are a popular choice for larger installations where space is not a constraint, such as solar farms or commercial rooftops. Their cost-effectiveness makes them an attractive option for projects with budget constraints.

Thin-film solar panels represent a different approach to solar technology. Unlike crystalline panels, thin-film panels are made by depositing photovoltaic material onto a substrate, such as glass, plastic, or metal. This process allows for the creation of flexible and lightweight panels that can be used in a variety of applications. Thin-film panels are less efficient than their crystalline counterparts, with efficiencies typically between 10% and 12%. However, their flexibility and adaptability make them ideal for unconventional installations, such as building-integrated photovoltaics (BIPV) or portable solar chargers. They are also less affected by shading and high temperatures, which can be advantageous in certain environments.

Bifacial solar panels are an innovative technology that captures sunlight on both sides of the panel. By utilizing reflected sunlight from the ground or surrounding surfaces, bifacial panels can increase energy production by up to 30% compared to traditional panels. These panels are often used in ground-mounted installations where the albedo effect, or the reflectivity of the ground, can be maximized. Bifacial panels are particularly effective in snowy or sandy environments, where the ground's reflectivity is high. Their ability to generate more electricity from the same footprint makes them an attractive option for maximizing energy yield.

Concentrated photovoltaic (CPV) panels take a different approach by using lenses or mirrors to focus sunlight onto a small, highly efficient solar cell. This concentration of sunlight allows CPV panels to achieve efficiencies exceeding 40%, making them the most efficient solar technology available. However, CPV systems require direct sunlight and precise tracking systems to maintain optimal alignment with the sun. As a result, they are best suited for regions with high solar insolation and clear skies. CPV technology is often used in utility-scale solar power plants, where the high efficiency can offset the additional complexity and cost of the system.

The choice of solar panel type depends on various factors, including the specific application, budget, available space, and environmental conditions. For residential installations, monocrystalline panels are often preferred for their high efficiency and aesthetic appeal. In contrast, polycrystalline panels may be chosen for larger commercial projects where cost is a primary consideration. Thin-film panels offer unique advantages for applications requiring flexibility or integration into building materials. Bifacial panels provide an opportunity to maximize energy production in suitable environments, while CPV panels offer unparalleled efficiency for large-scale solar farms in sunny regions.

Each type of solar panel has its own set of applications and considerations, and selecting the right one requires a careful assessment of the project's goals and constraints. By understanding the strengths and limitations of each technology, individuals and organizations can make informed decisions that align with their energy needs and sustainability objectives. As solar technology continues to evolve, new innovations and improvements will further expand the possibilities for harnessing the power of the sun, driving the transition to a cleaner and more sustainable energy future.

Installation and Maintenance of Solar Systems

Installing a solar power system is a transformative step towards energy independence and sustainability. The process involves several critical stages, each requiring careful planning and execution to ensure optimal performance and longevity. From site assessment to system maintenance, understanding the intricacies of solar installation can empower individuals and businesses to harness the sun's energy effectively.

The journey begins with a comprehensive site assessment, a crucial step that determines the feasibility and potential efficiency of a solar installation. This assessment involves evaluating the location's solar exposure, which is influenced by factors such as geographic latitude, roof orientation, and shading from nearby structures or vegetation. Tools like solar pathfinders or digital apps can help visualize the sun's path and identify potential obstructions. A thorough site assessment also considers the structural integrity of the roof or ground area where the panels will be installed, ensuring it can support the weight and withstand environmental conditions.

Once the site is deemed suitable, the next step is system design. This involves selecting the appropriate type and number of solar panels, inverters, and other components to meet the energy needs of the property. The design process takes into account the available space, budget, and energy consumption patterns. For residential installations, the goal is often to offset a significant portion of the household's electricity usage, while commercial systems may aim for larger-scale energy production. The design must also comply with local building codes and regulations, which can vary significantly between regions.

With the design finalized, the installation process can commence. This typically begins with securing the mounting system, which serves as the foundation for the solar panels. Roof-mounted systems are common for residential installations, using rails or brackets to attach the panels securely. Ground-mounted systems, often used in larger installations, require the construction of a sturdy frame anchored to the ground. The mounting system must be installed at the correct angle and orientation to maximize solar exposure throughout the year.

After the mounting system is in place, the solar panels are installed. This involves carefully positioning each panel and securing it to the mounting structure. The panels are then wired together in series or parallel configurations, depending on the desired voltage and current output. Proper wiring is essential to ensure the system operates safely and efficiently. The electrical connections must be weatherproofed to prevent moisture ingress, which can lead to corrosion and electrical faults.

The installation of the inverter is another critical step. The inverter converts the direct current (DC) generated by the solar panels into alternating current (AC), which is compatible with the electrical grid and household appliances. Inverters can be installed as a single unit for the entire system or as microinverters attached to each panel. The choice between these options depends on factors such as system size, shading, and budget. The inverter must be connected to the property's electrical panel, allowing the solar-generated electricity to be used on-site or fed back into the grid.

Once the system is installed, it undergoes a series of tests and inspections to ensure it meets safety and performance standards. This includes verifying the electrical connections, checking for proper grounding, and confirming the system's output matches the design specifications. Many regions require a final inspection by a local authority or utility company before the system can be activated.

Maintenance is a vital aspect of solar system ownership, ensuring the system continues to operate efficiently over its lifespan. Regular maintenance involves cleaning the solar panels to remove dirt, dust, and debris that can reduce their efficiency. In areas with frequent rain, natural cleaning may suffice, but in drier climates, periodic manual cleaning may be necessary. It's important to use appropriate cleaning tools and techniques to avoid damaging the panels.

In addition to cleaning, routine inspections of the system's components are essential. This includes checking the mounting structure for signs of wear or corrosion, inspecting the wiring for damage or loose connections, and ensuring the inverter is functioning correctly. Many modern inverters come with monitoring systems that provide real-time data on the system's performance, allowing for early detection of issues.

While solar systems are generally low-maintenance, occasional repairs may be needed. Common issues include inverter malfunctions, wiring faults, or damage from extreme weather events. Having a maintenance plan in place and working with a qualified solar technician can help address these issues promptly, minimizing downtime and maintaining energy production.

The installation and maintenance of solar systems require a blend of technical knowledge, practical skills, and attention to detail. By understanding the process and committing to regular upkeep, solar system owners can enjoy the benefits of renewable energy for decades. As technology advances and solar adoption grows, the practices and standards for installation and maintenance will continue to evolve, further enhancing the reliability and efficiency of solar power systems.

Economic and Environmental Impacts

The transition to renewable energy sources is not just a technological shift but a profound economic and environmental transformation. Solar power, as a leading renewable energy source, plays a pivotal role in this transition, offering significant economic benefits and environmental advantages that are reshaping the global energy landscape.

Economically, solar power presents a compelling case for investment and growth. The cost of solar technology has plummeted over the past decade, driven by advancements in manufacturing processes, increased competition, and economies of scale. This reduction in cost has made solar power one of the most affordable sources of electricity, rivaling traditional fossil fuels. For individuals and businesses, investing in solar power systems can lead to substantial savings on energy bills. By generating their own electricity, solar adopters can reduce or even eliminate their reliance on grid-supplied power, insulating themselves from fluctuating energy prices.

The economic impact of solar power extends beyond individual savings. The solar industry has become a significant driver of job creation, offering employment opportunities across various sectors, from manufacturing and installation to research and development. As the demand for solar energy continues to grow, so does the need for skilled workers, contributing to economic development and workforce diversification. In many regions, government incentives and tax credits further enhance the financial attractiveness of solar investments, encouraging more widespread adoption and stimulating local economies.

On a larger scale, solar power contributes to energy security and independence. By diversifying the energy mix and reducing reliance on imported fossil fuels, countries can enhance their energy resilience and reduce vulnerability to geopolitical tensions and supply disruptions. This shift towards domestic energy production can lead to a more stable and predictable energy market, benefiting both consumers and businesses.

Environmentally, the benefits of solar power are equally compelling. As a clean and renewable energy source, solar power generates electricity without emitting greenhouse gases or air pollutants. This characteristic makes it a crucial tool in the fight against climate change, helping to reduce carbon footprints and mitigate the impacts of global warming. By displacing fossil fuel-based energy sources, solar power can significantly decrease air pollution, leading to improved public health outcomes and reduced healthcare costs.

The environmental advantages of solar power extend to water conservation as well. Unlike traditional power plants, which often require large quantities of water for cooling, solar power systems operate with minimal water usage. This feature is particularly beneficial in arid regions or areas facing water scarcity, where conserving water resources is a critical concern.

Solar power also offers a sustainable solution for land use. While large-scale solar farms require significant land areas, they can be strategically located on marginal or degraded lands, minimizing competition with agricultural or ecologically sensitive areas. Additionally, innovative approaches such as agrivoltaics, which combine solar panels with agricultural activities, allow for dual land use, maximizing the productivity of available land.

Despite these benefits, the widespread adoption of solar power is not without challenges. The intermittent nature of solar energy, dependent on sunlight availability, necessitates the development of energy storage solutions and grid management strategies to ensure a reliable power supply. Advances in battery technology and smart grid systems are addressing these challenges, enhancing the integration of solar power into existing energy infrastructures.

The lifecycle of solar panels, from production to disposal, also presents environmental considerations. The manufacturing process involves the use of raw materials and energy, contributing to the overall environmental footprint of solar technology. However, ongoing research and innovation are focused on improving the sustainability of solar panel production, including the development of recycling programs to manage end-of-life panels responsibly.

The economic and environmental impacts of solar power are interconnected, influencing policy decisions and shaping the future of energy systems worldwide. As governments and organizations recognize the dual benefits of solar energy, they are increasingly incorporating it into their sustainability strategies and climate action plans. This alignment of economic and environmental goals is driving the transition towards a more sustainable and resilient energy future.

In conclusion, solar power offers a transformative solution to the dual challenges of economic growth and environmental sustainability. By harnessing the sun's energy, we can achieve significant economic benefits, from cost savings and job creation to energy security and market stability. At the same time, solar power provides a clean and renewable energy source that reduces greenhouse gas emissions, conserves water, and promotes sustainable land use. As we continue to innovate and expand the reach of solar technology, its economic and environmental impacts will play a crucial role in shaping a sustainable future for generations to come.

Case Studies: Successful Solar Projects

Solar energy has proven its potential through numerous successful projects worldwide, each demonstrating the versatility and impact of harnessing the sun's power. These case studies offer valuable insights into the practical applications of solar technology, showcasing how it can be tailored to meet diverse energy needs and environmental goals.

One of the most notable examples is the Noor Abu Dhabi solar plant in the United Arab Emirates. As one of the largest single-site solar projects globally, it spans over 8 square kilometers and boasts a capacity of 1.17 gigawatts. This massive installation is a testament to the scalability of solar power, providing clean energy to approximately 90,000 people and reducing carbon emissions by nearly 1 million metric tons annually. The success of Noor Abu Dhabi underscores the potential for solar energy to contribute significantly to national energy grids, particularly in regions with abundant sunlight.

In the United States, the Solar Star project in California stands out as a pioneering effort in utility-scale solar power. With a capacity of 579 megawatts, it was once the largest solar farm in the world. Solar Star's success lies in its innovative use of photovoltaic technology and its strategic location in the sunny Antelope Valley. The project not only supplies electricity to hundreds of thousands of homes but also serves as a model for integrating solar power into existing energy infrastructures. Its development has spurred economic growth in the region, creating jobs and fostering a local solar industry.

Australia's commitment to renewable energy is exemplified by the Hornsdale Power Reserve, which combines solar power with battery storage. Located in South Australia, this project integrates a 100-megawatt battery system with a nearby wind farm, demonstrating the potential for hybrid renewable energy solutions. The Hornsdale Power Reserve has proven instrumental in stabilizing the local grid, providing backup power during outages, and reducing electricity costs. Its success highlights the importance of energy storage in maximizing the benefits of solar power, ensuring a reliable and resilient energy supply.

In India, the Kamuthi Solar Power Project showcases the rapid deployment and scalability of solar technology in emerging markets. Situated in Tamil Nadu, this solar park covers an area of 10 square kilometers and has a capacity of 648 megawatts. Kamuthi's development was completed in a record time of just eight months, reflecting India's commitment to expanding its renewable energy capacity. The project plays a crucial role in meeting the country's growing energy demands while reducing its reliance on fossil fuels. It also serves as a catalyst for further solar investments in the region, contributing to India's ambitious renewable energy targets.

The success of solar projects is not limited to large-scale installations. In rural Bangladesh, the Solar Home Systems (SHS) initiative has transformed the lives of millions by providing off-grid solar solutions to households without access to electricity. Through microfinancing and community engagement, the SHS program has installed over 4 million solar systems, bringing light and power to remote areas. This initiative demonstrates the social and economic benefits of decentralized solar energy, empowering communities, improving quality of life, and fostering sustainable development.

In Europe, the Cestas Solar Park in France exemplifies the integration of solar power into agricultural landscapes. This 300-megawatt solar farm is located on agricultural land, demonstrating how solar energy can coexist with traditional land uses. The Cestas project has been designed to minimize environmental impact, incorporating measures to protect local biodiversity and maintain soil health. Its success highlights the potential for solar power to contribute to sustainable land management, offering a model for balancing energy production with environmental stewardship.

These case studies illustrate the diverse applications and benefits of solar energy across different contexts and scales. From large-scale power plants to community-based initiatives, solar projects are driving the transition to a sustainable energy future. Each project offers lessons in innovation, collaboration, and adaptability, showcasing the potential for solar power to address global energy challenges.

The success of these projects is underpinned by several key factors. Strategic site selection, leveraging areas with high solar insolation, is crucial for maximizing energy production. Technological innovation, such as the integration of energy storage and hybrid systems, enhances the reliability and efficiency of solar installations. Policy support and financial incentives play a vital role in facilitating project development, reducing costs, and encouraging investment. Community engagement and stakeholder collaboration ensure that projects meet local needs and gain public support.

As solar technology continues to evolve, new opportunities and challenges will emerge. The lessons learned from successful solar projects provide a foundation for future developments, guiding the expansion of solar energy in diverse settings. By building on these successes, we can accelerate the transition to a cleaner, more sustainable energy landscape, harnessing the power of the sun to meet the world's energy needs.

Chapter 2: Wind Energy: Capturing Nature's Power

Fundamentals of Wind Energy

Harnessing the power of the wind has been a pursuit of humanity for centuries, from the sails of ancient ships to the windmills of the agricultural revolution. Today, wind energy stands as a pillar of the renewable energy sector, offering a clean and sustainable alternative to fossil fuels. Understanding the fundamentals of wind energy involves exploring the principles of wind dynamics, the technology behind wind turbines, and the integration of wind power into modern energy systems.

At the heart of wind energy is the kinetic energy of moving air masses. Wind is generated by the uneven heating of the Earth's surface by the sun, creating pressure differences that drive air movement. This natural phenomenon is influenced by factors such as the Earth's rotation, topography, and atmospheric conditions. The potential for wind energy generation is determined by wind speed, which varies with location, altitude, and time of year. Areas with consistent and strong winds, such as coastal regions and open plains, are ideal for wind energy projects.

Wind turbines are the primary technology used to convert wind energy into electricity. These machines consist of several key components: the rotor, which includes the blades; the nacelle, housing the generator and gearbox; and the tower, which elevates the rotor to capture higher wind speeds. As wind flows over the blades, it creates lift, causing the rotor to spin. This rotational motion is transferred through the gearbox to the generator, where it is converted into electrical energy. The efficiency of a wind turbine is influenced by factors such as blade design, rotor size, and the height of the tower.

Modern wind turbines are marvels of engineering, designed to maximize energy capture while minimizing environmental impact. Advances in materials science and aerodynamics have led to the development of larger and more efficient turbines, capable of generating significant amounts of electricity. Offshore wind farms, which take advantage of stronger and more consistent winds over the ocean, represent a growing segment of the industry. These installations require specialized technology to withstand harsh marine conditions, but they offer the potential for substantial energy production.

The integration of wind energy into the electrical grid presents both opportunities and challenges. Wind power is inherently variable, as it depends on weather conditions that can change rapidly. This variability requires grid operators to balance wind energy with other power sources to ensure a stable and reliable electricity supply. Energy storage solutions, such as batteries and pumped hydro storage, play a crucial role in managing this variability, allowing excess wind energy to be stored and used when needed.

The economic and environmental benefits of wind energy are significant. Wind power is one of the most cost-effective sources of new electricity, with prices continuing to decline as technology advances and economies of scale are realized. The operation of wind turbines produces no greenhouse gas emissions, making wind energy a key component of efforts to combat climate change. Additionally, wind energy projects can stimulate local economies by creating jobs in manufacturing, installation, and maintenance.

Despite its advantages, wind energy development faces several challenges. Public opposition to wind farms can arise due to concerns about visual impact, noise, and effects on wildlife. Careful site selection and community engagement are essential to address these concerns and ensure the successful implementation of wind projects. Technological innovations, such as quieter turbine designs and wildlife-friendly technologies, are helping to mitigate some of these issues.

The future of wind energy is bright, with continued advancements in technology and increasing global commitment to renewable energy. Floating wind turbines, which can be deployed in deep waters where traditional fixed-bottom turbines are not feasible, represent a promising frontier. These innovations have the potential to unlock vast new areas for wind energy development, further expanding the reach and impact of this renewable resource.

In conclusion, wind energy is a dynamic and rapidly evolving field, offering a sustainable solution to the world's energy needs. By understanding the principles of wind dynamics, the technology of wind turbines, and the integration of wind power into energy systems, we can harness the power of the wind to create a cleaner and more sustainable future. As we continue to innovate and expand the use of wind energy, it will play an increasingly important role in the global transition to renewable energy.

Wind Turbine Technology and Design

Wind turbine technology has evolved dramatically over the past few decades, transforming from rudimentary designs into sophisticated machines capable of harnessing the wind's energy with remarkable efficiency. Understanding the intricacies of wind turbine technology and design is essential for appreciating how these structures convert kinetic energy into electricity and contribute to the global renewable energy landscape.

At the core of wind turbine technology is the rotor, which consists of blades that capture the wind's energy. The design of these blades is crucial, as it directly impacts the turbine's efficiency and performance. Modern wind turbine blades are typically made from composite materials such as fiberglass or carbon fiber, which offer a balance of strength, flexibility, and lightweight properties. The aerodynamic shape of the blades is inspired by the wings of an airplane, designed to create lift as wind passes over them. This lift causes the rotor to spin, converting the wind's kinetic energy into mechanical energy.

The rotor is connected to a shaft that transfers the mechanical energy to the nacelle, the central hub of the wind turbine. Inside the nacelle, a gearbox increases the rotational speed of the shaft, optimizing it for electricity generation. The gearbox is a critical component, as it ensures that the generator operates at its most efficient speed, even when wind speeds vary. Some modern wind turbines use direct-drive systems that eliminate the need for a gearbox, reducing mechanical complexity and maintenance requirements.

The generator, housed within the nacelle, is where mechanical energy is converted into electrical energy. Most wind turbines use either synchronous or asynchronous generators, each with its own advantages. Synchronous generators are often used in larger turbines due to their ability to maintain a constant frequency, while asynchronous generators are favored for their simplicity and cost-effectiveness. The choice of generator depends on factors such as turbine size, location, and grid connection requirements.

The tower of a wind turbine plays a vital role in its design, elevating the rotor and nacelle to capture higher wind speeds. Towers are typically constructed from steel or concrete and can reach heights of over 100 meters. The height of the tower is a critical design consideration, as wind speeds generally increase with altitude. Taller towers can access more consistent and powerful winds, enhancing the turbine's energy output. However, the increased height also presents engineering challenges, requiring robust structural design to withstand the forces exerted by the wind.

Wind turbine technology also includes sophisticated control systems that optimize performance and ensure safety. These systems monitor wind conditions and adjust the turbine's orientation and blade pitch to maximize energy capture. The yaw system rotates the nacelle to face the wind, while the pitch control system adjusts the angle of the blades to regulate rotational speed and prevent damage during high winds. Advanced sensors and software algorithms enable real-time monitoring and control, allowing turbines to operate efficiently across a range of wind conditions.

Offshore wind turbines represent a significant advancement in wind energy technology, designed to harness the stronger and more consistent winds found over the ocean. These turbines are typically larger and more powerful than their onshore counterparts, with capacities exceeding 10 megawatts. Offshore turbine design must account for the harsh marine environment, incorporating corrosion-resistant materials and robust foundations to withstand waves and saltwater exposure. Floating wind turbines are an emerging technology that allows for deployment in deeper waters, expanding the potential for offshore wind energy development.

The design of wind farms, whether onshore or offshore, involves careful planning and optimization. Turbines must be strategically spaced to minimize wake effects, where the turbulence created by one turbine reduces the efficiency of those downwind. Computational modeling and wind resource assessment are used to determine the optimal layout, balancing energy production with environmental and logistical considerations.

Wind turbine technology continues to advance, driven by the pursuit of greater efficiency, reliability, and cost-effectiveness. Innovations such as taller towers, longer blades, and advanced materials are pushing the boundaries of what is possible, enabling turbines to capture more energy from the wind. Research into new generator technologies, energy storage solutions, and grid integration strategies is further enhancing the role of wind energy in the global energy mix.

The future of wind turbine design holds exciting possibilities, with concepts such as airborne wind energy systems and vertical-axis turbines offering new approaches to capturing wind energy. These innovations have the potential to unlock new markets and applications, making wind energy accessible to a wider range of locations and communities.

In summary, wind turbine technology and design are at the forefront of the renewable energy revolution, offering a sustainable solution to the world's energy needs. By understanding the principles and components of wind turbines, we can appreciate the engineering marvels that harness the power of the wind and contribute to a cleaner, more sustainable future. As technology continues to evolve, wind energy will play an increasingly vital role in the transition to renewable energy, driving innovation and progress in the quest for a sustainable energy landscape.

Site Selection and Wind Farm Development

Choosing the right location for a wind farm is a critical step in the development process, as it directly influences the project's success and efficiency. The site selection process involves a comprehensive evaluation of various factors, including wind resources, environmental impact, land use, and logistical considerations. Each of these elements plays a vital role in determining the feasibility and potential output of a wind energy project.

The first and foremost consideration in site selection is the availability of wind resources. A thorough wind resource assessment is conducted to measure wind speed and direction over time, typically using anemometers and wind vanes installed at different heights. This data helps identify areas with consistent and strong winds, which are essential for maximizing energy production. Wind maps and meteorological models are also used to predict wind patterns and assess the long-term viability of a site. Coastal regions, open plains, and elevated areas are often prime candidates for wind farm development due to their favorable wind conditions.

Environmental impact is another crucial factor in site selection. Wind farms must be designed to minimize their ecological footprint and avoid disrupting local wildlife and habitats. Environmental assessments are conducted to evaluate potential impacts on bird and bat populations, as well as other sensitive species. Measures such as turbine placement, operational adjustments, and habitat conservation plans can mitigate these impacts. Additionally, the visual and noise impact of wind turbines on nearby communities must be considered, with efforts made to engage stakeholders and address any concerns.

Land use and ownership are important considerations in the site selection process. Wind farms require large areas of land, and securing access to suitable sites can be a complex process. Developers must negotiate land leases or purchase agreements with landowners, ensuring that the terms are favorable for both parties. In some cases, wind farms can coexist with existing land uses, such as agriculture or grazing, providing additional income streams for landowners. The compatibility of wind energy with other land uses is a key factor in gaining community support and facilitating project development.

Logistical considerations, such as proximity to transmission infrastructure and accessibility, also play a significant role in site selection. Wind farms must be connected to the electrical grid to deliver the generated power to consumers. The cost and feasibility of building transmission lines can significantly impact the overall project budget. Sites located near existing transmission infrastructure are often preferred, as they reduce the need for extensive new construction. Additionally, the accessibility of the site for construction and maintenance activities must be evaluated, taking into account road access, terrain, and weather conditions.

Once a suitable site is identified, the development process begins with securing the necessary permits and approvals. This involves navigating a complex regulatory landscape, which can vary significantly between regions and countries. Developers must comply with local, state, and federal regulations, which may include environmental permits, zoning approvals, and grid connection agreements. Engaging with regulatory authorities early in the process can help streamline the permitting process and address any potential issues.

Community engagement is a critical component of wind farm development. Building strong relationships with local communities and stakeholders is essential for gaining support and addressing concerns. Public consultations, informational meetings, and open communication channels can help build trust and foster collaboration. Developers should be transparent about the project's benefits and impacts, providing opportunities for community input and involvement. Community support can be a decisive factor in the success of a wind farm, influencing both the permitting process and long-term project viability.

The construction phase of wind farm development involves several key activities, including site preparation, turbine installation, and grid connection. Site preparation may involve clearing vegetation, grading land, and constructing access roads. Turbine installation requires specialized equipment and skilled labor, with each turbine assembled and erected on-site. The grid connection process involves installing electrical infrastructure, such as substations and transmission lines, to integrate the wind farm into the existing energy network.

Once construction is complete, the wind farm enters the operational phase, where it begins generating electricity. Ongoing maintenance and monitoring are essential to ensure optimal performance and longevity. Regular inspections, component replacements, and software updates are part of routine maintenance activities. Advanced monitoring systems provide real-time data on turbine performance, enabling operators to identify and address issues promptly.

The development of a wind farm is a complex and multifaceted process, requiring careful planning, collaboration, and execution. By selecting the right site and addressing key considerations, developers can create successful wind energy projects that contribute to a sustainable energy future. As technology advances and the demand for renewable energy grows, the principles of site selection and wind farm development will continue to evolve, driving innovation and progress in the wind energy sector.

Challenges and Solutions in Wind Energy

Wind energy has emerged as a cornerstone of the global transition to renewable energy, yet its development and deployment are not without challenges. These challenges span technical, environmental, economic, and social dimensions, each requiring innovative solutions to ensure the continued growth and success of wind energy projects.

One of the primary technical challenges in wind energy is the variability and intermittency of wind resources. Unlike conventional power plants that provide a steady output, wind energy production fluctuates with wind speed and weather conditions. This variability can pose challenges for grid operators who must balance supply and demand to maintain a stable electricity grid. To address this, advancements in energy storage technologies, such as batteries and pumped hydro storage, are being developed to store excess wind energy and release it when needed. Additionally, improved forecasting techniques using meteorological data and machine learning algorithms are enhancing the accuracy of wind predictions, allowing for better grid management and integration.

The environmental impact of wind farms is another significant concern. While wind energy is a clean and renewable source, the construction and operation of wind turbines can affect local ecosystems and wildlife. Birds and bats, in particular, are at risk of collision with turbine blades. To mitigate these impacts, developers are employing strategies such as careful site selection, turbine curtailment during peak migration periods, and the use of radar and acoustic deterrents to keep wildlife away from turbines. Ongoing research into turbine design and operation aims to further reduce these environmental impacts, ensuring that wind energy remains a sustainable option.

Economic challenges also play a role in the development of wind energy projects. The initial capital costs for wind farms can be substantial, including expenses for land acquisition, turbine manufacturing, and infrastructure development. While the cost of wind energy has decreased significantly over the past decade, financing remains a critical hurdle for many projects. Innovative financing models, such as power purchase agreements (PPAs) and green bonds, are helping to attract investment by providing stable revenue streams and reducing financial risk. Government incentives and subsidies also play a crucial role in making wind energy projects economically viable, encouraging private sector participation and fostering industry growth.

Social acceptance and community engagement are vital for the success of wind energy projects. Public opposition can arise due to concerns about visual impact, noise, and perceived effects on property values. To address these concerns, developers are increasingly involving local communities in the planning process, ensuring transparency and open communication. Community benefit schemes, such as revenue sharing or local investment opportunities, can also help build support by demonstrating the tangible benefits of wind energy projects. By fostering positive relationships with stakeholders, developers can overcome social barriers and gain the necessary approvals for project implementation.

Technological innovation continues to drive solutions to many of the challenges facing wind energy. The development of larger and more efficient turbines, capable of capturing more energy from the wind, is reducing the cost per megawatt-hour of electricity generated. Offshore wind technology, which harnesses the stronger and more consistent winds over the ocean, is expanding the potential for wind energy development. Floating wind turbines, in particular, offer the possibility of deploying wind farms in deeper waters, opening up new areas for exploration and reducing competition for land use.

Grid integration remains a critical challenge as the share of wind energy in the electricity mix increases. The variability of wind power requires grid operators to adapt their systems to accommodate fluctuating inputs. Smart grid technologies, which use digital communication and automation to optimize grid performance, are playing a key role in facilitating this integration. By enabling real-time monitoring and control, smart grids can enhance the reliability and efficiency of electricity distribution, ensuring that wind energy contributes effectively to the overall energy supply.

The future of wind energy is promising, with continued advancements in technology and growing global commitment to renewable energy. By addressing the challenges and implementing innovative solutions, the wind energy sector can continue to expand and play a pivotal role in the transition to a sustainable energy future. As we navigate these challenges, collaboration between industry, government, and communities will be essential to unlocking the full potential of wind energy and achieving a cleaner, more resilient energy system.

Future Prospects and Innovations

The horizon of wind energy is brimming with potential, driven by technological advancements and a global commitment to sustainable energy solutions. As the world seeks to reduce its carbon footprint and transition to cleaner energy sources, wind energy stands poised to play a pivotal role. The future prospects and innovations in this field promise to enhance efficiency, expand capacity, and integrate wind power more seamlessly into the global energy landscape.

One of the most exciting developments in wind energy is the advent of larger and more efficient wind turbines. Engineers are pushing the boundaries of turbine design, creating machines with longer blades and taller towers that can capture more energy from the wind. These advancements are not merely incremental; they represent a significant leap in the capacity of individual turbines, with some models now capable of generating over 15 megawatts of power. The increase in size and efficiency reduces the cost per megawatt-hour, making wind energy more competitive with traditional fossil fuels.

Offshore wind energy is another area ripe for innovation. The open seas offer stronger and more consistent winds than those found on land, providing an ideal environment for wind farms. However, the challenges of constructing and maintaining turbines in deep waters have historically limited offshore development. Enter floating wind turbines, a groundbreaking technology that allows turbines to be anchored in deeper waters where fixed-bottom structures are impractical. These floating platforms open up vast new areas for wind energy development, particularly in regions with deep coastal waters. As this technology matures, it promises to significantly expand the global capacity for offshore wind energy.

The integration of wind energy into existing power grids is a critical challenge that future innovations aim to address. The variability of wind power requires sophisticated grid management solutions to ensure a stable and reliable electricity supply. Smart grid technologies, which leverage digital communication and automation, are at the forefront of this effort. By enabling real-time monitoring and control of energy flows, smart grids can optimize the distribution of wind energy, balancing supply and demand more effectively. Additionally, advancements in energy storage, such as high-capacity batteries and other innovative storage solutions, are crucial for smoothing out the fluctuations in wind energy production.

Hybrid energy systems represent another promising avenue for the future of wind energy. By combining wind power with other renewable sources, such as solar or hydroelectric energy, these systems can provide a more consistent and reliable energy supply. Hybrid systems can take advantage of the complementary nature of different energy sources, using solar power during the day and wind power at night, for example. This approach not only enhances the reliability of renewable energy but also maximizes the use of available resources, reducing the need for fossil fuel backup.

The role of digital technology in wind energy is expanding, with innovations in data analytics, machine learning, and the Internet of Things (IoT) offering new ways to optimize performance and maintenance. Predictive analytics can forecast maintenance needs, reducing downtime and extending the lifespan of turbines. IoT devices can monitor turbine performance in real-time, providing valuable data that can be used to fine-tune operations and improve efficiency. These digital tools are transforming the way wind farms are managed, making them more efficient and cost-effective.

As wind energy technology continues to evolve, the industry is also exploring new materials and manufacturing techniques to improve turbine performance and reduce costs. Advanced materials, such as carbon fiber composites, offer greater strength and durability while being lighter than traditional materials. Additive manufacturing, or 3D printing, is being used to produce complex turbine components with greater precision and less waste. These innovations in materials and manufacturing are helping to drive down the cost of wind energy, making it more accessible and attractive to a wider range of markets.

The future of wind energy is not just about technological advancements; it also involves policy and regulatory frameworks that support the growth of the industry. Governments around the world are recognizing the importance of renewable energy in achieving climate goals and are implementing policies to encourage investment in wind energy. These policies include subsidies, tax incentives, and renewable energy mandates that create a favorable environment for wind energy development. As these frameworks evolve, they will play a crucial role in shaping the future of the industry.

Public perception and community engagement are also key factors in the future success of wind energy. As wind farms become more prevalent, it is essential to address concerns about visual impact, noise, and effects on local wildlife. Engaging with communities and stakeholders early in the planning process can help build support and address potential issues. By demonstrating the economic and environmental benefits of wind energy, developers can foster positive relationships with communities and ensure the long-term success of their projects.

The future prospects and innovations in wind energy are vast and varied, offering exciting opportunities for growth and development. As technology advances and the global commitment to renewable energy strengthens, wind energy will continue to play a vital role in the transition to a sustainable energy future. By embracing these innovations and addressing the challenges ahead, the wind energy industry can unlock its full potential and contribute to a cleaner, more resilient energy system for generations to come.

Principles of Hydropower Generation

Hydropower, one of the oldest and most reliable forms of renewable energy, harnesses the energy of flowing water to generate electricity. Its principles are rooted in the natural water cycle, where the sun's energy drives evaporation, precipitation, and the flow of water through rivers and streams. Understanding the principles of hydropower generation involves exploring the mechanics of converting kinetic and potential energy into electrical power, the various types of hydropower systems, and the environmental and social considerations associated with this energy source.

At the core of hydropower generation is the conversion of water's kinetic and potential energy into mechanical energy, which is then transformed into electricity. This process begins with the movement of water, typically from a higher elevation to a lower one, driven by gravity. The potential energy of water stored at a height is converted into kinetic energy as it flows downward. This kinetic energy is captured by turbines, which are mechanical devices designed to rotate as water flows over or through them. The rotation of the turbine's blades turns a shaft connected to a generator, where mechanical energy is converted into electrical energy through electromagnetic induction.

There are several types of hydropower systems, each suited to different geographical and hydrological conditions. The most common type is the conventional dam-based hydropower system, where a dam is constructed to create a reservoir that stores water. This stored water is released in a controlled manner to drive turbines located at the base of the dam. The height of the water column, known as the "head," and the flow rate determine the amount of energy that can be generated. Dams not only facilitate energy generation but also provide benefits such as flood control, irrigation, and water supply.

Run-of-river hydropower systems, in contrast, do not require large reservoirs. Instead, they divert a portion of a river's flow through a channel or penstock to drive turbines. These systems are typically smaller in scale and have a lower environmental impact, as they do not significantly alter the natural flow of the river. However, their energy output can be more variable, depending on seasonal changes in river flow.

Pumped storage hydropower is a unique type of system that acts as both a generator and a storage facility. During periods of low electricity demand, excess energy from the grid is used to pump water from a lower reservoir to an upper reservoir. When demand increases, the stored water is released back to the lower reservoir, driving turbines and generating electricity. This system provides a valuable means of balancing supply and demand, offering grid stability and energy storage capabilities.

The efficiency of hydropower systems is influenced by several factors, including the design of the turbines, the head and flow rate of the water, and the overall system configuration. Turbines are selected based on the specific characteristics of the site, with common types including Kaplan, Francis, and Pelton turbines. Kaplan turbines are suitable for low-head, high-flow applications, while Francis turbines are versatile and can operate efficiently across a range of head and flow conditions.

Pelton turbines are ideal for high-head, low-flow scenarios, often found in mountainous regions.

While hydropower offers numerous benefits, including low greenhouse gas emissions and reliable energy production, it also presents environmental and social challenges. The construction of dams and reservoirs can disrupt local ecosystems, affecting fish populations and altering natural river flows. To mitigate these impacts, environmental assessments are conducted to evaluate potential effects on biodiversity and water quality. Fish ladders and bypass systems are often implemented to facilitate the migration of aquatic species, while reservoir management practices aim to maintain ecological balance.

Social considerations are also important in hydropower development, particularly in regions where communities rely on rivers for their livelihoods. The displacement of communities due to reservoir creation can lead to social and economic challenges. Engaging with affected communities and stakeholders is essential to address concerns and ensure that the benefits of hydropower projects are shared equitably. Compensation, resettlement programs, and community development initiatives are critical components of responsible hydropower development.

The future of hydropower generation is shaped by technological advancements and a growing emphasis on sustainability. Innovations in turbine design and materials are enhancing the efficiency and environmental performance of hydropower systems. Small-scale and micro-hydropower technologies are expanding access to renewable energy in remote and off-grid areas, providing decentralized energy solutions that empower local communities.

As the global energy landscape evolves, hydropower remains a vital component of the renewable energy mix. Its ability to provide baseload power, support grid stability, and offer energy storage solutions makes it a valuable asset in the transition to a sustainable energy future. By understanding the principles of hydropower generation and addressing the associated challenges, we can harness the power of water to create a cleaner, more resilient energy system for generations to come.

Types of Hydropower Plants

Hydropower plants, a cornerstone of renewable energy, come in various forms, each tailored to specific geographical and hydrological conditions. Understanding the different types of hydropower plants is essential for appreciating how they harness the energy of flowing water to generate electricity. These plants are categorized based on their design, operation, and the way they utilize water resources, offering diverse solutions to meet energy demands.

The most common type of hydropower plant is the impoundment facility, which involves the construction of a dam across a river to create a reservoir. This reservoir stores water, providing a controlled and reliable source of energy. The potential energy of the stored water is converted into kinetic energy as it is released through turbines located at the base of the dam. Impoundment plants can generate a consistent supply of electricity, making them ideal for meeting baseload energy demands. They also offer additional benefits such as flood control, irrigation, and recreational opportunities. However, the construction of large dams can have significant environmental and social impacts, including habitat disruption and community displacement.

Run-of-river hydropower plants, in contrast, do not require large reservoirs. Instead, they divert a portion of a river's flow through a channel or penstock to drive turbines. These plants rely on the natural flow of the river, with minimal storage capacity. As a result, their energy output can vary with seasonal changes in river flow. Run-of-river plants are typically smaller in scale and have a lower environmental impact compared to impoundment facilities, as they do not significantly alter the natural flow of the river. They are well-suited for regions with consistent river flow and are often used to complement other forms of energy generation.

Pumped storage hydropower plants serve a dual purpose, acting as both a generator and a storage facility. These plants consist of two reservoirs at different elevations. During periods of low electricity demand, excess energy from the grid is used to pump water from the lower reservoir to the upper reservoir. When demand increases, the stored water is released back to the lower reservoir, driving turbines and generating electricity. Pumped storage plants provide a valuable means of balancing supply and demand, offering grid stability and energy storage capabilities. They are particularly useful for integrating variable renewable energy sources, such as wind and solar, into the grid.

Another innovative type of hydropower plant is the tidal power facility, which harnesses the energy of ocean tides to generate electricity. Tidal power plants are typically located in coastal areas with significant tidal ranges. They use the rise and fall of tides to drive turbines, either through the construction of a tidal barrage or the deployment of underwater turbines in tidal streams. Tidal power is a predictable and reliable source of energy, as tidal cycles are driven by the gravitational pull of the moon and sun. However, the development of tidal power plants can be challenging due to the harsh marine environment and potential impacts on marine ecosystems.

Micro-hydropower plants represent a smaller-scale approach to hydropower generation, often used in remote or off-grid areas. These plants typically generate less than 100 kilowatts of electricity, making them suitable for individual homes, small communities, or agricultural operations. Micro-hydropower systems can be installed in small streams or rivers, with minimal infrastructure and environmental impact. They offer a decentralized energy solution, empowering local communities and providing access to renewable energy in areas where larger plants are not feasible.

The choice of hydropower plant type depends on various factors, including the availability of water resources, geographical conditions, and energy needs. Each type of plant offers unique advantages and challenges, requiring careful consideration and planning. As technology advances and the demand for renewable energy grows, the diversity of hydropower plants will continue to expand, offering innovative solutions to meet the world's energy needs.

Hydropower remains a vital component of the global energy mix, providing a reliable and sustainable source of electricity. By understanding the different types of hydropower plants and their applications, we can harness the power of water to create a cleaner, more resilient energy system for future generations.

As we explore new technologies and approaches, hydropower will continue to play a crucial role in the transition to a sustainable energy future.

Environmental Considerations and Mitigation

The development of renewable energy sources, such as hydropower, is essential for reducing our reliance on fossil fuels and mitigating climate change. However, the construction and operation of hydropower plants can have significant environmental impacts that must be carefully considered and managed. Understanding these impacts and implementing effective mitigation strategies is crucial for ensuring that hydropower remains a sustainable and environmentally responsible energy source.

One of the primary environmental concerns associated with hydropower is the alteration of natural river ecosystems. The construction of dams and reservoirs can disrupt the natural flow of rivers, affecting water quality, sediment transport, and aquatic habitats. These changes can have cascading effects on the entire ecosystem, impacting fish populations, plant life, and other wildlife. To mitigate these impacts, environmental assessments are conducted to evaluate the potential effects of hydropower projects on local ecosystems. These assessments inform the design and operation of the plant, guiding decisions on dam placement, reservoir size, and water flow management.

Fish populations are particularly vulnerable to the impacts of hydropower development. Dams can obstruct migratory routes, preventing fish from reaching spawning grounds and reducing their populations. To address this issue, fish ladders and bypass systems are often implemented to provide safe passage for fish around dams. These structures are designed to mimic natural river conditions, allowing fish to navigate upstream and downstream without harm. Additionally, turbine design and operation can be optimized to minimize the risk of injury to fish passing through the plant.

The creation of reservoirs can also lead to the inundation of large areas of land, affecting terrestrial habitats and displacing wildlife. This can result in the loss of biodiversity and the disruption of local ecosystems. To mitigate these impacts, careful site selection and reservoir management practices are essential. By choosing sites with minimal ecological value and implementing measures to protect and restore habitats, developers can reduce the environmental footprint of hydropower projects. Reservoir management practices, such as maintaining water levels and flow rates that mimic natural conditions, can also help preserve aquatic and terrestrial ecosystems.

Water quality is another important consideration in hydropower development. The construction and operation of dams can affect the temperature, oxygen levels, and nutrient content of water, impacting aquatic life and downstream ecosystems. To mitigate these effects, water quality monitoring programs are implemented to track changes and identify potential issues. Adaptive management strategies, such as adjusting water releases and implementing aeration systems, can help maintain water quality and support healthy ecosystems.

The social and cultural impacts of hydropower development are also significant. The creation of reservoirs can lead to the displacement of communities and the loss of cultural heritage sites. Engaging with affected communities and stakeholders is essential to address these concerns and ensure that the benefits of hydropower projects are shared equitably. Compensation, resettlement programs, and community development initiatives are critical components of responsible hydropower development. By involving local communities in the planning process and providing opportunities for participation and input, developers can build trust and foster positive relationships.

Climate change considerations are increasingly important in the planning and operation of hydropower projects. Changes in precipitation patterns, temperature, and river flow can affect the availability of water resources and the performance of hydropower plants. To address these challenges, developers are incorporating climate resilience into project design and operation. This includes using climate models to predict future water availability, designing flexible and adaptive infrastructure, and implementing water management strategies that account for changing conditions.

Innovations in technology and design are also playing a role in reducing the environmental impact of hydropower. Advances in turbine technology, such as fish-friendly turbines and variable-speed generators, are helping to minimize harm to aquatic life and improve efficiency. The use of advanced materials and construction techniques can reduce the environmental footprint of hydropower projects, making them more sustainable and cost-effective.

The integration of environmental considerations into the planning and operation of hydropower projects is essential for ensuring their long-term sustainability. By understanding and addressing the potential impacts on ecosystems, wildlife, and communities, developers can create hydropower projects that balance energy production with environmental stewardship. As the demand for renewable energy continues to grow, the principles of environmental consideration and mitigation will play a crucial role in shaping the future of hydropower and ensuring its contribution to a sustainable energy future.

Economic Viability and Community Impact

The economic viability of hydropower projects is a critical factor in their development and success. Hydropower, as a renewable energy source, offers numerous benefits, including low operational costs and long-term sustainability. However, the initial capital investment required for constructing hydropower facilities can be substantial. Understanding the economic aspects of hydropower and its impact on local communities is essential for ensuring that these projects are both financially feasible and socially beneficial.

The financial viability of a hydropower project hinges on several key factors, including the cost of construction, the availability of financing, and the projected revenue from electricity sales. The construction of a hydropower plant involves significant expenses, including land acquisition, engineering and design, equipment procurement, and labor. These costs can vary widely depending on the size and type of the project, as well as the geographical and environmental conditions of the site. Despite the high upfront costs, hydropower plants have relatively low operating and maintenance expenses, making them an attractive long-term investment.

Securing financing for hydropower projects is a crucial step in their development. Investors and financial institutions are often drawn to the stability and predictability of hydropower as an energy source. Power purchase agreements (PPAs) are commonly used to provide a guaranteed revenue stream, reducing financial risk and attracting investment. These agreements involve a long-term contract between the hydropower developer and an electricity buyer, typically a utility company, ensuring a fixed price for the electricity generated. Government incentives, such as tax credits and subsidies, can also play a significant role in enhancing the economic viability of hydropower projects by reducing the overall cost and encouraging private sector participation.

The economic benefits of hydropower extend beyond the project itself, contributing to local and regional development. Hydropower projects can create jobs during the construction and operational phases, providing employment opportunities for local communities. The development of infrastructure, such as roads and bridges, can improve access to remote areas and stimulate economic growth. Additionally, hydropower projects can generate revenue for local governments through taxes and royalties, supporting public services and community development initiatives.

While the economic benefits of hydropower are significant, it is essential to consider the potential impacts on local communities. The construction of dams and reservoirs can lead to the displacement of communities and the loss of agricultural land, affecting livelihoods and cultural heritage. Engaging with affected communities and stakeholders is crucial to address these concerns and ensure that the benefits of hydropower projects are shared equitably. Compensation and resettlement programs are critical components of responsible hydropower development, providing support for displaced communities and helping them rebuild their lives.

Community engagement and participation are vital for the success of hydropower projects. By involving local communities in the planning and decision-making process, developers can build trust and foster positive relationships. Transparent communication and open dialogue can help address concerns and ensure that community needs and priorities are considered. Community benefit schemes, such as revenue sharing or local investment opportunities, can also help build support by demonstrating the tangible benefits of hydropower projects.

The social and economic impacts of hydropower projects can vary widely depending on the specific context and location. In some cases, hydropower development can lead to significant improvements in living standards and quality of life, providing access to electricity, clean water, and improved infrastructure. In other cases, the impacts may be more mixed, with both positive and negative effects on local communities. Understanding and addressing these impacts is essential for ensuring that hydropower projects contribute to sustainable development and social well-being.

The economic viability and community impact of hydropower projects are closely intertwined, requiring a balanced approach that considers both financial and social factors. By understanding the economic aspects of hydropower and engaging with local communities, developers can create projects that are both financially successful and socially responsible. As the demand for renewable energy continues to grow, the principles of economic viability and community impact will play a crucial role in shaping the future of hydropower and ensuring its contribution to a sustainable energy future.

Innovations in Small-Scale Hydropower

Small-scale hydropower is emerging as a vital component of the renewable energy landscape, offering flexible and sustainable solutions for energy generation in diverse settings. Unlike large-scale hydropower projects, which often require significant infrastructure and can have substantial environmental impacts, small-scale hydropower systems are designed to be more adaptable and environmentally friendly. These systems are particularly well-suited for remote or off-grid areas, providing reliable energy access where traditional power infrastructure may be lacking.

One of the most significant innovations in small-scale hydropower is the development of modular and portable systems. These systems are designed for easy installation and operation, allowing them to be deployed quickly and efficiently in a variety of settings. Modular systems can be scaled up or down to meet specific energy needs, making them ideal for communities with fluctuating demand. Portable systems, on the other hand, offer the flexibility to be moved and reinstalled as needed, providing a versatile solution for temporary or seasonal energy requirements.

The use of advanced materials and manufacturing techniques is also driving innovation in small-scale hydropower. Lightweight and durable materials, such as composites and advanced polymers, are being used to construct turbines and other components, reducing the overall weight and cost of the systems. Additive manufacturing, or 3D printing, is enabling the production of complex components with greater precision and less waste, further enhancing the efficiency and affordability of small-scale hydropower systems.

Innovations in turbine design are playing a crucial role in improving the performance and efficiency of small-scale hydropower systems. Traditional turbine designs, such as Kaplan and Francis turbines, are being adapted for smaller applications, while new designs, such as cross-flow and Archimedes screw turbines, are being developed specifically for low-head and low-flow conditions. These innovative designs allow small-scale hydropower systems to operate efficiently in a wider range of environments, including small streams, canals, and even urban settings.

The integration of digital technology is transforming the way small-scale hydropower systems are monitored and managed. Smart sensors and IoT devices are being used to collect real-time data on system performance, water flow, and environmental conditions. This data can be analyzed to optimize system operation, predict maintenance needs, and improve overall efficiency. Remote monitoring and control capabilities also allow operators to manage systems from a distance, reducing the need for on-site personnel and lowering operational costs.

Hybrid energy systems, which combine small-scale hydropower with other renewable energy sources, are another area of innovation. By integrating hydropower with solar, wind, or biomass energy, these systems can provide a more consistent and reliable energy supply. Hybrid systems can take advantage of the complementary nature of different energy sources, using solar power during the day and hydropower at night, for example. This approach not only enhances the reliability of renewable energy but also maximizes the use of available resources, reducing the need for fossil fuel backup.

The environmental benefits of small-scale hydropower are significant, as these systems typically have a lower impact on local ecosystems compared to large-scale projects. By using existing water infrastructure, such as irrigation canals or water supply systems, small-scale hydropower can generate energy without the need for new dams or reservoirs. This minimizes habitat disruption and preserves natural river flows, supporting biodiversity and ecosystem health.

Community involvement and local ownership are key factors in the success of small-scale hydropower projects. By engaging with local communities and stakeholders, developers can ensure that projects are designed to meet specific needs and priorities. Community ownership models, where local residents have a stake in the project, can enhance support and foster a sense of ownership and responsibility. These models also provide economic benefits, as revenue generated from energy sales can be reinvested in local development initiatives.

The potential for small-scale hydropower to contribute to sustainable development is immense. By providing reliable and affordable energy access, these systems can support economic growth, improve living standards, and enhance resilience to climate change. As technology continues to advance and the demand for renewable energy grows, small-scale hydropower will play an increasingly important role in the global energy transition.

The future of small-scale hydropower is bright, with ongoing innovations and advancements paving the way for more efficient, cost-effective, and environmentally friendly solutions. By embracing these innovations and addressing the challenges ahead, small-scale hydropower can unlock its full potential and contribute to a cleaner, more resilient energy system for generations to come.

Chapter 4: Biomass and Bioenergy: Organic Power Sources

Understanding Biomass Energy

Biomass energy, a versatile and renewable energy source, has been utilized by humans for centuries. It involves the conversion of organic materials, such as plant and animal waste, into usable energy. This form of energy is derived from the sun, as plants capture solar energy through photosynthesis, storing it in the form of chemical energy. When biomass is burned or converted into biofuels, this stored energy is released, providing heat, electricity, or fuel for transportation.

The primary sources of biomass energy include agricultural residues, forestry by-products, animal manure, and organic waste from industries and households. Agricultural residues, such as corn stalks, rice husks, and wheat straw, are abundant and readily available. These materials can be collected after harvest and used as feedstock for biomass energy production. Forestry by-products, including wood chips, sawdust, and bark, are generated during logging and wood processing activities. These materials can be used directly as fuel or processed into pellets for more efficient combustion.

Animal manure, a by-product of livestock farming, is another valuable source of biomass energy. It can be processed through anaerobic digestion, a biological process that breaks down organic matter in the absence of oxygen, producing biogas. Biogas, primarily composed of methane and carbon dioxide, can be used for heating, electricity generation, or as a vehicle fuel. Organic waste from industries and households, such as food scraps and yard trimmings, can also be converted into energy through anaerobic digestion or composting.

The conversion of biomass into energy can be achieved through various technologies, each with its own advantages and applications. Combustion is the most straightforward method, involving the direct burning of biomass to produce heat. This heat can be used for cooking, space heating, or generating steam for electricity production. Modern biomass combustion systems are designed to be highly efficient and environmentally friendly, with advanced emission control technologies to minimize air pollution.

Gasification is another technology used to convert biomass into energy. It involves heating biomass in a low-oxygen environment to produce a combustible gas mixture known as syngas. Syngas can be used for electricity generation, heating, or as a feedstock for producing chemicals and fuels. Gasification offers several advantages, including higher efficiency and the ability to process a wide range of feedstocks.

Pyrolysis is a thermal conversion process that decomposes biomass at high temperatures in the absence of oxygen, producing bio-oil, syngas, and biochar. Bio-oil can be refined into transportation fuels or used as a chemical feedstock, while biochar can be used as a soil amendment to improve soil fertility and sequester carbon. Pyrolysis is a flexible technology that can be adapted to different feedstocks and end-use applications.

Biochemical conversion processes, such as fermentation and anaerobic digestion, are used to produce biofuels from biomass. Fermentation involves the conversion of sugars and starches into ethanol, a renewable fuel that can be blended with gasoline or used as a standalone fuel. Anaerobic digestion, as mentioned earlier, produces biogas from organic waste, providing a renewable alternative to natural gas.

The environmental benefits of biomass energy are significant, as it offers a renewable and carbon-neutral alternative to fossil fuels. When biomass is used for energy, the carbon dioxide released during combustion is offset by the carbon dioxide absorbed by plants during their growth. This closed carbon cycle helps reduce greenhouse gas emissions and mitigate climate change. Additionally, biomass energy can help reduce waste and promote sustainable land management practices.

Despite its advantages, biomass energy also presents challenges that must be addressed to ensure its sustainability. The collection and transportation of biomass feedstocks can be resource-intensive and costly, particularly in remote or rural areas. Ensuring a consistent and reliable supply of feedstock is essential for the economic viability of biomass energy projects. Land use and competition with food production are also important considerations, as the cultivation of energy crops can impact food security and biodiversity.

Technological advancements and innovation are driving the development of more efficient and sustainable biomass energy systems. Research is focused on improving feedstock conversion technologies, enhancing the efficiency of biomass combustion and gasification systems, and developing new biofuels and bioproducts. The integration of biomass energy with other renewable energy sources, such as solar and wind, is also being explored to create hybrid systems that offer greater reliability and flexibility.

Biomass energy has the potential to play a significant role in the transition to a sustainable energy future. By understanding the principles and technologies behind biomass energy, we can harness its potential to provide clean, renewable energy while supporting economic development and environmental stewardship. As we continue to innovate and address the challenges ahead, biomass energy will remain a vital component of the global energy landscape, contributing to a more sustainable and resilient energy system for generations to come.

Conversion Technologies and Processes

Conversion technologies and processes are at the heart of transforming raw materials into usable energy, playing a pivotal role in the renewable energy landscape. These technologies encompass a wide range of methods that convert various forms of biomass, solar, wind, and other renewable resources into electricity, heat, or fuel. Understanding these processes is essential for harnessing the full potential of renewable energy sources and advancing towards a sustainable energy future.

Biomass conversion technologies are diverse, reflecting the variety of feedstocks and end-use applications. Combustion is one of the oldest and most straightforward methods, involving the direct burning of biomass to produce heat. This heat can be used for residential heating, industrial processes, or to generate steam for electricity production. Modern combustion systems are designed to maximize efficiency and minimize emissions, incorporating advanced technologies such as fluidized bed combustion and gasification.

Gasification is a more sophisticated conversion process that involves heating biomass in a low-oxygen environment to produce syngas, a mixture of carbon monoxide, hydrogen, and methane. Syngas can be used for electricity generation, heating, or as a feedstock for producing chemicals and fuels. Gasification offers several advantages over traditional combustion, including higher efficiency and the ability to process a wide range of feedstocks, from agricultural residues to municipal solid waste.

Pyrolysis is another thermal conversion process that decomposes biomass at high temperatures in the absence of oxygen, producing bio-oil, syngas, and biochar. Bio-oil can be refined into transportation fuels or used as a chemical feedstock, while biochar can be applied to soils to improve fertility and sequester carbon. Pyrolysis is a flexible technology that can be adapted to different feedstocks and end-use applications, offering a sustainable solution for waste management and energy production.

Biochemical conversion processes, such as fermentation and anaerobic digestion, are used to produce biofuels from biomass. Fermentation involves the conversion of sugars and starches into ethanol, a renewable fuel that can be blended with gasoline or used as a standalone fuel. This process is widely used in the production of bioethanol from crops like corn and sugarcane. Anaerobic digestion, on the other hand, breaks down organic matter in the absence of oxygen to produce biogas, a renewable alternative to natural gas. Biogas can be used for heating, electricity generation, or as a vehicle fuel, providing a versatile and sustainable energy source.

Solar energy conversion technologies harness the power of the sun to generate electricity or heat. Photovoltaic (PV) systems convert sunlight directly into electricity using semiconductor materials, such as silicon, that exhibit the photovoltaic effect. PV systems are highly scalable, ranging from small rooftop installations to large utility-scale solar farms. Advances in PV technology, such as the development of thin-film solar cells and bifacial modules, are improving efficiency and reducing costs, making solar energy more accessible and affordable.

Concentrated solar power (CSP) systems use mirrors or lenses to focus sunlight onto a small area, generating heat that can be used to produce steam and drive a turbine for electricity generation. CSP systems are particularly well-suited for regions with high solar insolation and can be integrated with thermal energy storage to provide a stable and reliable power supply. Innovations in CSP technology, such as the use of molten salt as a heat transfer fluid, are enhancing efficiency and enabling longer periods of energy storage.

Wind energy conversion technologies capture the kinetic energy of wind and convert it into electricity using wind turbines. Modern wind turbines are highly efficient and can operate in a wide range of wind conditions, from gentle breezes to strong gusts. Advances in turbine design, such as larger rotor diameters and taller towers, are increasing energy capture and reducing the cost of wind energy. Offshore wind farms, which take advantage of stronger and more consistent winds over the ocean, are expanding the potential for wind energy generation.

Hydropower conversion technologies harness the energy of flowing water to generate electricity. Impoundment facilities, which involve the construction of a dam to create a reservoir, are the most common type of hydropower plant. These facilities store water and release it through turbines to generate electricity, providing a reliable and consistent power supply. Run-of-river systems, which divert a portion of a river's flow through a channel or penstock, offer a more environmentally friendly alternative, with minimal impact on natural river ecosystems.

Geothermal energy conversion technologies tap into the heat stored beneath the Earth's surface to generate electricity or provide direct heating. Geothermal power plants use steam or hot water from underground reservoirs to drive turbines and produce electricity. Direct-use applications, such as district heating and greenhouse heating, utilize geothermal heat for residential, commercial, and agricultural purposes. Advances in geothermal technology, such as enhanced geothermal systems (EGS), are expanding the potential for geothermal energy development in regions with lower natural geothermal activity.

The integration of conversion technologies with energy storage systems is enhancing the reliability and flexibility of renewable energy sources. Energy storage technologies, such as batteries, pumped hydro storage, and thermal storage, allow excess energy to be stored and used when demand is high or when renewable generation is low. This integration is critical for balancing supply and demand, ensuring grid stability, and maximizing the use of renewable resources.

Conversion technologies and processes are at the forefront of the renewable energy revolution, offering innovative solutions for sustainable energy production. By understanding and advancing these technologies, we can unlock the full potential of renewable energy sources, reduce our reliance on fossil fuels, and create a cleaner, more resilient energy system for future generations. As we continue to innovate and address the challenges ahead, conversion technologies will play a pivotal role in shaping the future of energy and driving the transition to a sustainable energy future.

Sustainability and Environmental Impact

Sustainability and environmental impact are central considerations in the development and deployment of renewable energy technologies. As the world grapples with the challenges of climate change and environmental degradation, the transition to sustainable energy sources has become imperative. Renewable energy technologies, such as solar, wind, biomass, and hydropower, offer the promise of reducing greenhouse gas emissions and minimizing the environmental footprint of energy production. However, it is essential to carefully assess and manage the environmental impacts of these technologies to ensure that they contribute positively to sustainability goals.

The concept of sustainability in the context of energy refers to the ability to meet present energy needs without compromising the ability of future generations to meet their own needs. This involves balancing economic, social, and environmental considerations to create an energy system that is resilient, equitable, and environmentally responsible. Renewable energy technologies play a crucial role in achieving sustainability by providing clean, abundant, and renewable sources of energy that can reduce reliance on fossil fuels and mitigate climate change.

Solar energy is one of the most sustainable forms of energy, as it harnesses the sun's abundant and inexhaustible energy.

Photovoltaic (PV) systems, which convert sunlight directly into electricity, have a relatively low environmental impact compared to conventional energy sources. The production of solar panels does involve the use of raw materials and energy, but advances in manufacturing processes and recycling technologies are reducing the environmental footprint of PV systems. Additionally, solar energy systems have no emissions during operation, making them a clean and sustainable energy source.

Wind energy is another highly sustainable form of energy, as it captures the kinetic energy of wind to generate electricity. Wind turbines have a low environmental impact during operation, with no emissions or water use. However, the construction and installation of wind farms can have localized environmental impacts, such as habitat disruption and noise. Careful site selection and community engagement are essential to minimize these impacts and ensure that wind energy projects are developed in a sustainable manner.

Biomass energy, derived from organic materials such as plant and animal waste, offers a renewable and carbon-neutral alternative to fossil fuels. When biomass is used for energy, the carbon dioxide released during combustion is offset by the carbon dioxide absorbed by plants during their growth. This closed carbon cycle helps reduce greenhouse gas emissions and mitigate climate change. However, the sustainability of biomass energy depends on the responsible management of feedstocks and land use. Ensuring a consistent and sustainable supply of biomass feedstocks, while avoiding competition with food production and preserving biodiversity, is critical for the long-term sustainability of biomass energy.

Hydropower, which harnesses the energy of flowing water to generate electricity, is one of the oldest and most established forms of renewable energy. It offers a reliable and consistent power supply with low operational emissions. However, the construction of dams and reservoirs can have significant environmental impacts, including habitat disruption, changes in river flow, and displacement of communities. To ensure the sustainability of hydropower, it is essential to conduct thorough environmental assessments, engage with affected communities, and implement mitigation measures to minimize negative impacts.

The integration of renewable energy technologies with energy storage systems is enhancing the sustainability and reliability of renewable energy sources. Energy storage technologies, such as batteries and pumped hydro storage, allow excess energy to be stored and used when demand is high or when renewable generation is low. This integration is critical for balancing supply and demand, ensuring grid stability, and maximizing the use of renewable resources.

The environmental impact of renewable energy technologies extends beyond their direct emissions and resource use. It is important to consider the entire lifecycle of these technologies, from raw material extraction and manufacturing to installation, operation, and decommissioning. Lifecycle assessments (LCAs) are used to evaluate the environmental impacts of renewable energy technologies throughout their entire lifecycle, providing valuable insights into areas for improvement and optimization.

Community engagement and participation are essential for ensuring the sustainability and social acceptance of renewable energy projects. By involving local communities in the planning and decision-making process, developers can build trust and foster positive relationships. Transparent communication and open dialogue can help address concerns and ensure that community needs and priorities are considered. Community benefit schemes, such as revenue sharing or local investment opportunities, can also help build support by demonstrating the tangible benefits of renewable energy projects.

The transition to a sustainable energy future requires a holistic approach that considers the environmental, social, and economic dimensions of energy production and use. By understanding and addressing the environmental impacts of renewable energy technologies, we can harness their potential to provide clean, renewable energy while supporting economic development and environmental stewardship. As we continue to innovate and address the challenges ahead, sustainability and environmental impact will remain central to the development and deployment of renewable energy technologies, ensuring their contribution to a cleaner, more resilient energy system for generations to come.

Economic and Social Benefits

Economic and social benefits are integral to the adoption and expansion of renewable energy technologies. As the world shifts towards sustainable energy solutions, the positive impacts on economies and communities are becoming increasingly evident. Renewable energy not only addresses environmental concerns but also offers a pathway to economic growth, job creation, and social development.

The economic benefits of renewable energy are multifaceted, beginning with the potential for job creation. The renewable energy sector is labor-intensive, requiring a diverse range of skills for the design, installation, operation, and maintenance of energy systems. This demand for skilled labor translates into numerous job opportunities across various sectors, from engineering and manufacturing to construction and project management. As renewable energy projects proliferate, they stimulate local economies by creating jobs and supporting ancillary industries, such as transportation and logistics.

Investment in renewable energy infrastructure also drives economic growth by attracting capital and fostering innovation. Governments and private investors are increasingly recognizing the potential of renewable energy to deliver stable and long-term returns. This influx of investment not only supports the development of new projects but also encourages research and development in cutting-edge technologies. As a result, renewable energy becomes a catalyst for technological advancement, driving down costs and improving efficiency.

The decentralization of energy production, a hallmark of renewable energy systems, further enhances economic resilience. By reducing reliance on centralized power plants and fossil fuel imports, communities can achieve greater energy independence and security. This shift not only mitigates the risks associated with volatile energy markets but also empowers local economies to retain more of their energy expenditures. Decentralized energy systems, such as rooftop solar panels and community wind farms, enable individuals and communities to generate their own electricity, reducing utility bills and freeing up resources for other economic activities.

Renewable energy projects also contribute to infrastructure development, particularly in rural and underserved areas. The construction of renewable energy facilities often necessitates improvements in transportation, communication, and utility infrastructure. These enhancements can have a lasting impact on local communities, improving access to essential services and fostering economic development. Moreover, renewable energy projects can provide a reliable and affordable energy supply, supporting the growth of small businesses and industries that rely on consistent power.

The social benefits of renewable energy are equally compelling, as they contribute to improved quality of life and social equity. Access to clean and affordable energy is a fundamental driver of social development, enabling communities to improve health, education, and overall well-being. Renewable energy projects can provide electricity to remote and off-grid areas, where traditional energy infrastructure is lacking. This access to energy can transform lives, powering schools, healthcare facilities, and homes, and enabling the use of modern technologies that enhance productivity and quality of life.

Renewable energy also plays a crucial role in addressing energy poverty, a condition that affects millions of people worldwide. By providing affordable and sustainable energy solutions, renewable energy projects can help lift communities out of poverty and reduce inequalities. The deployment of renewable energy technologies can empower marginalized groups, such as women and indigenous communities, by creating opportunities for participation and leadership in the energy sector. Community-based renewable energy projects, in particular, can foster social cohesion and empowerment by involving local stakeholders in decision-making and ownership.

The environmental benefits of renewable energy, while primarily ecological, also have significant social implications. By reducing greenhouse gas emissions and air pollution, renewable energy projects contribute to improved public health and environmental quality. Cleaner air and water, resulting from reduced reliance on fossil fuels, can lead to lower rates of respiratory and cardiovascular diseases, benefiting individuals and communities. The preservation of natural ecosystems and biodiversity, supported by sustainable energy practices, enhances the quality of life for current and future generations.

Education and awareness are critical components of realizing the social benefits of renewable energy. By promoting understanding and acceptance of renewable energy technologies, communities can make informed decisions about their energy future. Educational initiatives, such as workshops, training programs, and public awareness campaigns, can empower individuals with the knowledge and skills needed to participate in the renewable energy transition. These efforts can also inspire the next generation of innovators and leaders in the energy sector, ensuring the continued growth and success of renewable energy.

The economic and social benefits of renewable energy are interconnected, creating a virtuous cycle of development and progress. As renewable energy technologies become more widespread, they drive economic growth, create jobs, and improve quality of life, while also addressing pressing environmental challenges. By embracing renewable energy, societies can build a more sustainable, equitable, and prosperous future for all. The transition to renewable energy is not only an environmental imperative but also an opportunity to transform economies and uplift communities, paving the way for a brighter and more sustainable future.

Case Studies: Biomass Success Stories

Biomass energy has emerged as a powerful tool in the quest for sustainable energy solutions, with numerous success stories demonstrating its potential to transform communities and industries. These case studies highlight the diverse applications of biomass energy, showcasing its ability to provide clean, reliable, and affordable energy while supporting economic development and environmental stewardship.

One notable success story comes from Sweden, a country renowned for its commitment to renewable energy. In the city of Växjö, biomass energy has played a pivotal role in the municipality's ambitious goal to become fossil fuel-free. Växjö's district heating system, which supplies heat to homes and businesses, is powered primarily by biomass sourced from local forestry residues. This system not only reduces greenhouse gas emissions but also supports the local economy by creating jobs in the forestry and energy sectors. The success of Växjö's biomass initiative has inspired other cities around the world to adopt similar approaches, demonstrating the scalability and effectiveness of biomass energy in urban settings.

In India, the use of biomass energy has been instrumental in addressing energy poverty and improving quality of life in rural communities. The Husk Power Systems project, based in Bihar, India, has developed an innovative model for converting rice husks, an abundant agricultural residue, into electricity. By installing small-scale biomass gasification plants in rural villages,

Husk Power Systems provides reliable and affordable electricity to communities that previously relied on kerosene lamps and diesel generators. This access to electricity has transformed daily life, enabling children to study after dark, improving healthcare services, and supporting local businesses. The success of Husk Power Systems has garnered international recognition and investment, paving the way for the expansion of biomass energy solutions in other regions.

In the United States, the use of biomass energy in the agricultural sector has demonstrated significant environmental and economic benefits. The Poet-DSM Advanced Biofuels project in Emmetsburg, Iowa, is a prime example of how agricultural residues can be harnessed to produce renewable energy. This facility converts corn stover, the leaves and stalks left after corn harvest, into cellulosic ethanol, a renewable transportation fuel. By utilizing agricultural residues, the project reduces waste, lowers greenhouse gas emissions, and provides an additional revenue stream for farmers. The success of the Poet-DSM project has spurred further investment in advanced biofuels, highlighting the potential of biomass energy to drive innovation and sustainability in the agricultural sector.

In Brazil, the sugarcane industry has long been a leader in the production of bioenergy, with sugarcane bagasse, the fibrous residue left after juice extraction, serving as a valuable feedstock for biomass energy. The RaízenEnergia project, a joint venture between Shell and Cosan, has taken this concept to new heights by developing one of the world's largest biomass power plants. Located in São Paulo, the facility uses sugarcane bagasse to generate electricity, supplying power to the national grid and reducing reliance on fossil fuels. This project not only showcases the potential of biomass energy to support large-scale power generation but also underscores the importance of integrating renewable energy into existing industrial processes.

In Kenya, the use of biomass energy has been instrumental in promoting sustainable development and environmental conservation. The Kenya Tea Development Agency (KTDA) has implemented a biomass energy project that utilizes tea factory waste, such as wood chips and sawdust, to generate heat and electricity for tea processing. This initiative reduces the reliance on fossil fuels, lowers production costs, and supports sustainable forestry practices. The success of the KTDA project has encouraged other industries in Kenya to explore biomass energy solutions, contributing to the country's efforts to achieve energy security and environmental sustainability.

In Austria, the Güssing Renewable Energy project has transformed a small town into a model of sustainable energy production. By utilizing locally sourced biomass, such as wood chips and agricultural residues, the project generates electricity, heat, and biofuels for the community. This integrated approach not only reduces greenhouse gas emissions but also supports local economic development by creating jobs and fostering innovation. The success of the Güssing project has attracted international attention, serving as a blueprint for other communities seeking to transition to renewable energy.

These case studies illustrate the diverse applications and benefits of biomass energy, highlighting its potential to drive sustainable development and address pressing environmental challenges. By harnessing locally available resources and engaging communities in the energy transition, biomass energy projects can deliver significant economic, social, and environmental benefits. As the world continues to seek sustainable energy solutions, the success stories of biomass energy offer valuable insights and inspiration for future initiatives. Through innovation, collaboration, and a commitment to sustainability, biomass energy can play a vital role in shaping a cleaner, more resilient energy future.

Chapter 5: Geothermal Energy: Tapping into Earth's Heat

Basics of Geothermal Energy

Geothermal energy, a powerful and sustainable resource, taps into the Earth's internal heat to provide a reliable and clean source of energy. This form of energy has been utilized for centuries, with ancient civilizations using hot springs for bathing and heating. Today, geothermal energy is harnessed for electricity generation, direct heating applications, and even cooling, offering a versatile solution to modern energy needs.

At its core, geothermal energy originates from the heat generated by the Earth's formation and the radioactive decay of minerals within its crust. This heat is stored in rocks and fluids beneath the Earth's surface, creating geothermal reservoirs. These reservoirs can be accessed through wells, allowing the heat to be extracted and utilized for various applications. The temperature and depth of these reservoirs determine their suitability for different uses, ranging from electricity generation to direct heating.

Electricity generation from geothermal energy is primarily achieved through geothermal power plants, which convert the Earth's heat into electrical power. There are three main types of geothermal power plants: dry steam, flash steam, and binary cycle. Each type is suited to specific geothermal conditions and employs different technologies to harness the Earth's heat.

Dry steam power plants are the oldest type of geothermal power plant, utilizing steam directly from geothermal reservoirs to drive turbines and generate electricity. These plants are typically located in areas with high-temperature geothermal resources, such as the Geysers in California. The simplicity and efficiency of dry steam plants make them an attractive option for regions with suitable geothermal conditions.

Flash steam power plants are the most common type of geothermal power plant, accounting for the majority of geothermal electricity generation worldwide. These plants use high-pressure hot water from geothermal reservoirs, which is brought to the surface and allowed to "flash" into steam as the pressure decreases. The steam is then used to drive turbines and generate electricity. Flash steam plants are well-suited to regions with moderate to high-temperature geothermal resources.

Binary cycle power plants are a newer and more versatile technology, capable of utilizing lower temperature geothermal resources. These plants use a secondary fluid with a lower boiling point than water, such as isobutane or pentane, to transfer heat from the geothermal fluid. The secondary fluid vaporizes and drives a turbine, generating electricity. Binary cycle plants are particularly advantageous in areas with lower temperature geothermal resources, as they can operate efficiently at temperatures as low as 85°C (185°F).

Beyond electricity generation, geothermal energy is widely used for direct heating applications, providing a sustainable and cost-effective alternative to fossil fuels. Direct use applications include district heating, greenhouse heating, aquaculture, and industrial processes. In district heating systems, geothermal heat is distributed through a network of pipes to provide space heating and hot water to residential, commercial, and industrial buildings. This approach is particularly popular in countries like Iceland, where abundant geothermal resources provide heating for the majority of the population.

Geothermal heat pumps, also known as ground-source heat pumps, offer another innovative application of geothermal energy. These systems use the stable temperature of the Earth's subsurface to provide heating, cooling, and hot water for buildings. By circulating a fluid through a series of underground pipes, geothermal heat pumps transfer heat between the building and the ground, providing efficient and sustainable climate control. This technology is suitable for a wide range of climates and can significantly reduce energy consumption and greenhouse gas emissions.

The environmental benefits of geothermal energy are significant, as it offers a low-emission and renewable alternative to fossil fuels. Geothermal power plants emit minimal greenhouse gases compared to conventional power plants, contributing to climate change mitigation. Additionally, geothermal energy has a small land footprint, as power plants and direct use facilities require relatively little space compared to other energy sources. This makes geothermal energy an attractive option for regions with limited land availability.

Despite its advantages, geothermal energy also presents challenges that must be addressed to ensure its sustainable development. The exploration and development of geothermal resources can be costly and time-consuming, requiring significant investment and expertise. Additionally, the availability of geothermal resources is geographically limited, with the most suitable sites often located in tectonically active regions. This can pose challenges for the widespread adoption of geothermal energy, particularly in areas without access to high-temperature resources.

Technological advancements and innovation are driving the development of more efficient and sustainable geothermal energy systems. Enhanced geothermal systems (EGS), for example, are an emerging technology that aims to expand the potential of geothermal energy by creating artificial reservoirs in areas with hot, dry rock. By injecting water into these reservoirs, EGS can generate steam and produce electricity, even in regions without natural geothermal reservoirs. This technology has the potential to significantly increase the availability of geothermal energy and reduce reliance on fossil fuels.

Geothermal energy has the potential to play a significant role in the transition to a sustainable energy future. By understanding the principles and technologies behind geothermal energy, we can harness its potential to provide clean, reliable, and renewable energy while supporting economic development and environmental stewardship. As we continue to innovate and address the challenges ahead, geothermal energy will remain a vital component of the global energy landscape, contributing to a more sustainable and resilient energy system for generations to come.

Types of Geothermal Systems

Geothermal energy, a cornerstone of renewable energy solutions, is harnessed through various systems designed to tap into the Earth's internal heat. These systems are tailored to the specific geothermal resources available, ranging from high-temperature steam fields to low-temperature ground heat. Understanding the different types of geothermal systems is crucial for effectively utilizing this abundant and sustainable energy source.

The most prominent type of geothermal system is the hydrothermal system, which exploits naturally occurring reservoirs of hot water and steam. These systems are typically found in regions with high geothermal gradients, such as volcanic areas and tectonic plate boundaries. Hydrothermal systems are further categorized into two main types: liquid-dominated and vapor-dominated systems.

Liquid-dominated systems are the most common type of hydrothermal system, characterized by reservoirs filled with hot water. These systems are typically accessed through wells that bring the hot water to the surface, where it can be used for electricity generation or direct heating applications. In electricity generation, the hot water is often flashed into steam to drive turbines, a process that is efficient and widely used in geothermal power plants around the world.

Vapor-dominated systems, on the other hand, are less common but highly efficient. These systems are characterized by reservoirs where steam is the dominant phase, allowing for direct use in electricity generation. The Geysers in California is a prime example of a vapor-dominated geothermal field, where steam is extracted and used to power turbines. The efficiency and simplicity of vapor-dominated systems make them an attractive option for electricity generation, although their occurrence is limited to specific geological settings.

Beyond hydrothermal systems, enhanced geothermal systems (EGS) represent a cutting-edge approach to geothermal energy extraction. EGS technology aims to expand the potential of geothermal energy by creating artificial reservoirs in areas with hot, dry rock. This is achieved by injecting water into the rock, creating fractures that allow the water to circulate and absorb heat. The heated water is then pumped back to the surface, where it can be used for electricity generation or direct heating. EGS has the potential to significantly increase the availability of geothermal energy, making it accessible in regions without natural hydrothermal resources.

Geothermal heat pumps, also known as ground-source heat pumps, offer another innovative application of geothermal energy. These systems utilize the stable temperature of the Earth's subsurface to provide heating, cooling, and hot water for buildings. By circulating a fluid through a series of underground pipes, geothermal heat pumps transfer heat between the building and the ground, providing efficient and sustainable climate control. This technology is suitable for a wide range of climates and can significantly reduce energy consumption and greenhouse gas emissions.

Direct use geothermal systems are designed to utilize geothermal heat without the need for electricity generation. These systems are employed in a variety of applications, including district heating, greenhouse heating, aquaculture, and industrial processes. In district heating systems, geothermal heat is distributed through a network of pipes to provide space heating and hot water to residential, commercial, and industrial buildings. This approach is particularly popular in countries like Iceland, where abundant geothermal resources provide heating for the majority of the population.

Geothermal desalination is an emerging application of geothermal energy, offering a sustainable solution to water scarcity. By using geothermal heat to power desalination processes, such as multi-effect distillation or reverse osmosis, freshwater can be produced from seawater or brackish water. This approach not only provides a reliable source of freshwater but also reduces the energy consumption and environmental impact associated with conventional desalination methods.

The environmental benefits of geothermal systems are significant, as they offer a low-emission and renewable alternative to fossil fuels. Geothermal power plants emit minimal greenhouse gases compared to conventional power plants, contributing to climate change mitigation. Additionally, geothermal systems have a small land footprint, as power plants and direct use facilities require relatively little space compared to other energy sources. This makes geothermal energy an attractive option for regions with limited land availability.

Despite their advantages, geothermal systems also present challenges that must be addressed to ensure their sustainable development. The exploration and development of geothermal resources can be costly and time-consuming, requiring significant investment and expertise. Additionally, the availability of geothermal resources is geographically limited, with the most suitable sites often located in tectonically active regions. This can pose challenges for the widespread adoption of geothermal energy, particularly in areas without access to high-temperature resources.

Technological advancements and innovation are driving the development of more efficient and sustainable geothermal systems. Research into new drilling techniques, reservoir management, and heat extraction methods is expanding the potential of geothermal energy and reducing costs. Collaboration between governments, industry, and research institutions is essential to overcome the challenges and unlock the full potential of geothermal systems.

Geothermal energy has the potential to play a significant role in the transition to a sustainable energy future. By understanding the different types of geothermal systems and their applications, we can harness their potential to provide clean, reliable, and renewable energy while supporting economic development and environmental stewardship. As we continue to innovate and address the challenges ahead, geothermal systems will remain a vital component of the global energy landscape, contributing to a more sustainable and resilient energy system for generations to come.

Exploration and Development Techniques

Exploring and developing geothermal resources requires a blend of scientific expertise, advanced technology, and strategic planning. The journey from identifying a potential geothermal site to harnessing its energy involves a series of meticulous steps, each crucial for ensuring the viability and sustainability of the project. Understanding these techniques is essential for anyone venturing into the realm of geothermal energy.

The exploration phase begins with a comprehensive assessment of the geological characteristics of a region. Geologists and geophysicists play a pivotal role in this stage, employing various methods to identify areas with high geothermal potential. One of the primary techniques used is geological mapping, which involves studying the surface features and rock formations to infer the subsurface conditions. This mapping provides valuable insights into the tectonic activity, fault lines, and volcanic history of the area, all of which are indicators of geothermal activity.

Geophysical surveys complement geological mapping by providing a deeper understanding of the subsurface conditions. These surveys utilize techniques such as seismic reflection, magnetotellurics, and gravity measurements to detect anomalies that may indicate the presence of geothermal reservoirs. Seismic reflection, for instance, involves sending sound waves into the ground and analyzing the reflected signals to create a detailed image of the subsurface structures. This method is particularly useful for identifying faults and fractures that can serve as conduits for geothermal fluids.

Geochemical analysis is another critical component of the exploration process. By sampling and analyzing the chemical composition of surface and subsurface fluids, scientists can infer the temperature and composition of the geothermal reservoir. This information is vital for assessing the potential energy output and determining the suitability of the site for development. Geochemical surveys often involve the collection of water and gas samples from hot springs, fumaroles, and wells, followed by laboratory analysis to identify key indicators such as silica, chloride, and isotopic ratios.

Once a promising site has been identified, the next step is to conduct exploratory drilling. This phase involves drilling a series of test wells to directly access the geothermal reservoir and gather data on its temperature, pressure, and fluid characteristics. Exploratory drilling is a critical and costly step, as it provides the most accurate information about the reservoir's potential. The data collected from these wells is used to create a detailed reservoir model, which informs the design and development of the geothermal power plant.

The development phase begins with the design and construction of production wells, which are used to extract geothermal fluids from the reservoir. These wells are typically drilled to depths ranging from a few hundred meters to several kilometers, depending on the depth and temperature of the reservoir. Advanced drilling techniques, such as directional drilling and hydraulic fracturing, are often employed to maximize the efficiency and output of the wells. Directional drilling allows for the precise targeting of specific reservoir zones, while hydraulic fracturing enhances the permeability of the rock, facilitating the flow of geothermal fluids.

In addition to production wells, reinjection wells are also constructed to return cooled geothermal fluids back into the reservoir. This practice is essential for maintaining reservoir pressure and ensuring the long-term sustainability of the geothermal resource. Reinjection also minimizes the environmental impact of geothermal operations by preventing the depletion of local water resources and reducing the risk of surface subsidence.

The construction of the geothermal power plant is the final step in the development process. The design of the plant is tailored to the specific characteristics of the geothermal resource, with considerations for the temperature, pressure, and chemical composition of the fluids. The plant typically includes turbines, generators, heat exchangers, and cooling systems, all of which work together to convert the geothermal heat into electricity. The choice of technology, whether it be dry steam, flash steam, or binary cycle, depends on the temperature and phase of the geothermal fluids.

Throughout the exploration and development process, environmental and social considerations are paramount. Environmental impact assessments (EIAs) are conducted to evaluate the potential effects of geothermal operations on the local ecosystem and communities. These assessments inform the implementation of mitigation measures, such as noise reduction, habitat restoration, and community engagement, to minimize negative impacts and ensure the project's sustainability.

Community involvement is also a critical aspect of successful geothermal development. Engaging with local stakeholders, including residents, businesses, and government agencies, fosters transparency and builds trust. By involving the community in decision-making processes and providing opportunities for local employment and investment, geothermal projects can deliver tangible benefits and gain social acceptance.

The exploration and development of geothermal resources are complex and multifaceted endeavors, requiring a combination of scientific expertise, technological innovation, and strategic planning. By understanding and implementing these techniques, developers can unlock the potential of geothermal energy, providing a sustainable and reliable source of power for generations to come. As the demand for clean energy continues to grow, the exploration and development of geothermal resources will play an increasingly important role in the global energy landscape, contributing to a more sustainable and resilient future.

Environmental and Economic Considerations

Balancing environmental and economic considerations is crucial in the development and deployment of geothermal energy projects. As a renewable energy source, geothermal energy offers significant advantages in terms of sustainability and low emissions. However, like any energy project, it requires careful planning and management to ensure that its benefits are maximized while minimizing potential negative impacts.

Geothermal energy is often lauded for its minimal environmental footprint compared to fossil fuels. One of its most significant advantages is the low level of greenhouse gas emissions associated with its use. Geothermal power plants emit a fraction of the carbon dioxide produced by coal or natural gas plants, making them an attractive option for reducing the carbon footprint of energy production. Additionally, geothermal energy is a reliable and consistent power source, unaffected by weather conditions, which enhances its appeal as a sustainable energy solution.

Despite these benefits, geothermal projects can pose environmental challenges that must be addressed. One of the primary concerns is the potential for land subsidence, which can occur when large volumes of geothermal fluids are extracted from underground reservoirs. This can lead to the sinking or settling of the ground surface, potentially damaging infrastructure and ecosystems. To mitigate this risk, developers often implement reinjection strategies, where cooled geothermal fluids are returned to the reservoir to maintain pressure and stability.

Water usage is another critical consideration in geothermal energy projects. While geothermal power plants generally use less water than fossil fuel plants, the extraction and reinjection processes can impact local water resources. Ensuring sustainable water management practices, such as using non-potable water sources and recycling fluids, is essential to minimize the impact on local water supplies and ecosystems.

The release of trace gases and minerals from geothermal fluids is another environmental consideration. While emissions are generally low, geothermal fluids can contain hydrogen sulfide, a gas with a distinct odor that can be harmful in high concentrations. Advanced scrubbing technologies and proper site management can effectively control these emissions, ensuring they remain within safe limits. Additionally, the disposal of mineral-laden brine, a byproduct of geothermal energy production, requires careful handling to prevent contamination of soil and water resources.

Biodiversity and habitat preservation are also important factors in the planning and development of geothermal projects. The construction and operation of geothermal facilities can disrupt local ecosystems, particularly in sensitive or protected areas. Conducting thorough environmental impact assessments (EIAs) and engaging with local stakeholders can help identify potential impacts and develop strategies to mitigate them. This might include habitat restoration, wildlife corridors, and other conservation measures to protect local flora and fauna.

On the economic front, geothermal energy presents both opportunities and challenges. The initial costs of geothermal exploration and development can be high, often requiring significant investment in drilling and infrastructure. However, once operational, geothermal power plants typically have low operating and maintenance costs, providing a stable and cost-effective energy source over the long term. The economic viability of geothermal projects is often enhanced by government incentives, such as tax credits and grants, which can offset initial capital expenditures and encourage investment.

Geothermal energy can also drive economic development by creating jobs and supporting local industries. The construction and operation of geothermal facilities require a diverse workforce, including engineers, geologists, technicians, and support staff. Additionally, the presence of a reliable and affordable energy source can attract businesses and industries to the area, further boosting economic growth. In regions with abundant geothermal resources, such as Iceland and parts of the United States, geothermal energy has become a cornerstone of the local economy, providing both energy security and economic resilience.

Community engagement and social acceptance are critical components of successful geothermal projects. Involving local communities in the planning and decision-making processes can build trust and support for the project. This might include public consultations, educational programs, and opportunities for local employment and investment. By demonstrating the benefits of geothermal energy and addressing community concerns, developers can foster positive relationships and ensure the long-term success of the project.

The integration of geothermal energy into existing energy systems also presents economic considerations. Geothermal power plants can provide baseload power, complementing intermittent renewable sources like wind and solar. This can enhance grid stability and reduce reliance on fossil fuels, contributing to a more sustainable and resilient energy system. However, integrating geothermal energy into the grid may require upgrades to transmission infrastructure and changes to regulatory frameworks, which can involve additional costs and complexities.

Innovation and technological advancements continue to drive the economic and environmental potential of geothermal energy. Research into new drilling techniques, reservoir management, and heat extraction methods is expanding the range of viable geothermal resources and reducing costs. Enhanced geothermal systems (EGS), for example, offer the potential to access geothermal energy in regions without natural hydrothermal reservoirs, significantly increasing the availability of this renewable resource.

Balancing environmental and economic considerations is essential for the sustainable development of geothermal energy. By understanding and addressing the potential impacts and benefits, developers can harness the full potential of geothermal energy, providing a clean, reliable, and cost-effective energy source for the future. As the global demand for renewable energy continues to grow, geothermal energy will play an increasingly important role in the transition to a sustainable energy system, contributing to a more resilient and prosperous future for all.

Innovations and Future Directions

Innovation is the lifeblood of the geothermal energy sector, driving advancements that expand its potential and address the challenges it faces. As the world seeks sustainable energy solutions, the geothermal industry is poised to play a pivotal role, thanks to cutting-edge technologies and forward-thinking strategies that promise to reshape its landscape.

One of the most promising innovations in geothermal energy is the development of enhanced geothermal systems (EGS). Unlike traditional geothermal systems that rely on naturally occurring hydrothermal resources, EGS technology creates artificial reservoirs in hot, dry rock formations. By injecting water into these formations, EGS can generate steam and produce electricity, even in regions without natural geothermal reservoirs. This breakthrough has the potential to significantly increase the availability of geothermal energy, making it accessible in areas previously considered unsuitable for development.

Advancements in drilling technology are also revolutionizing the geothermal industry. High-temperature drilling techniques, such as laser-assisted drilling and plasma drilling, are being explored to improve efficiency and reduce costs. These methods offer the potential to reach greater depths and access higher temperature resources, unlocking new geothermal reserves. Additionally, the use of advanced materials and coatings in drilling equipment is enhancing durability and performance, further driving down costs and expanding the feasibility of geothermal projects.

The integration of geothermal energy with other renewable energy sources is another area of innovation. Hybrid systems that combine geothermal with solar or wind power can provide a more stable and reliable energy supply, addressing the intermittency issues associated with some renewables. For example, solar-geothermal hybrid plants can use solar energy to boost the temperature of geothermal fluids, increasing the efficiency of electricity generation. These integrated systems offer a promising pathway to a more resilient and sustainable energy grid.

Geothermal energy storage is an emerging field that holds great promise for enhancing the flexibility and reliability of geothermal power. By using geothermal reservoirs as natural storage systems, excess energy can be stored as heat and released when needed, providing a consistent power supply. This approach not only maximizes the use of geothermal resources but also supports grid stability and reduces reliance on fossil fuels.

Digitalization and data analytics are transforming the way geothermal projects are managed and optimized. The use of sensors, remote monitoring, and data analytics allows for real-time tracking of geothermal operations, enabling more efficient resource management and maintenance. Predictive analytics can identify potential issues before they become critical, reducing downtime and improving the overall performance of geothermal facilities. This data-driven approach is enhancing the economic viability and sustainability of geothermal projects.

The exploration of unconventional geothermal resources is opening new frontiers for the industry. Supercritical geothermal systems, which exist at extremely high temperatures and pressures, offer the potential for significantly higher energy output compared to conventional systems. Research into accessing these resources is ongoing, with the potential to revolutionize the geothermal landscape and provide a substantial boost to global energy supplies.

Community engagement and social innovation are also playing a crucial role in the future of geothermal energy. By involving local communities in the planning and development of geothermal projects, developers can build trust and support for their initiatives. This might include educational programs, public consultations, and opportunities for local employment and investment. By demonstrating the benefits of geothermal energy and addressing community concerns, developers can foster positive relationships and ensure the long-term success of their projects.

The role of policy and regulation in shaping the future of geothermal energy cannot be overstated. Supportive policies, such as tax incentives, grants, and streamlined permitting processes, are essential for encouraging investment and innovation in the sector. Governments and regulatory bodies play a critical role in creating an enabling environment for geothermal development, ensuring that projects are economically viable and environmentally sustainable.

International collaboration and knowledge sharing are vital for advancing geothermal technology and expanding its global reach. By working together, countries can share best practices, research findings, and technological advancements, accelerating the development and deployment of geothermal energy worldwide. Collaborative initiatives, such as international research partnerships and industry conferences, provide valuable platforms for exchanging ideas and driving innovation.

The future of geothermal energy is bright, with innovations and advancements paving the way for its continued growth and success. By embracing new technologies, integrating with other renewable sources, and fostering community and policy support, the geothermal industry is well-positioned to play a leading role in the transition to a sustainable energy future. As we look ahead, the potential of geothermal energy to provide clean, reliable, and renewable power is more promising than ever, offering a pathway to a more resilient and prosperous world.

www.ingramcontent.com/pod-product-compliance
Lightning Source LLC
Chambersburg PA
CBHW072031150726
47999CB00002B/852